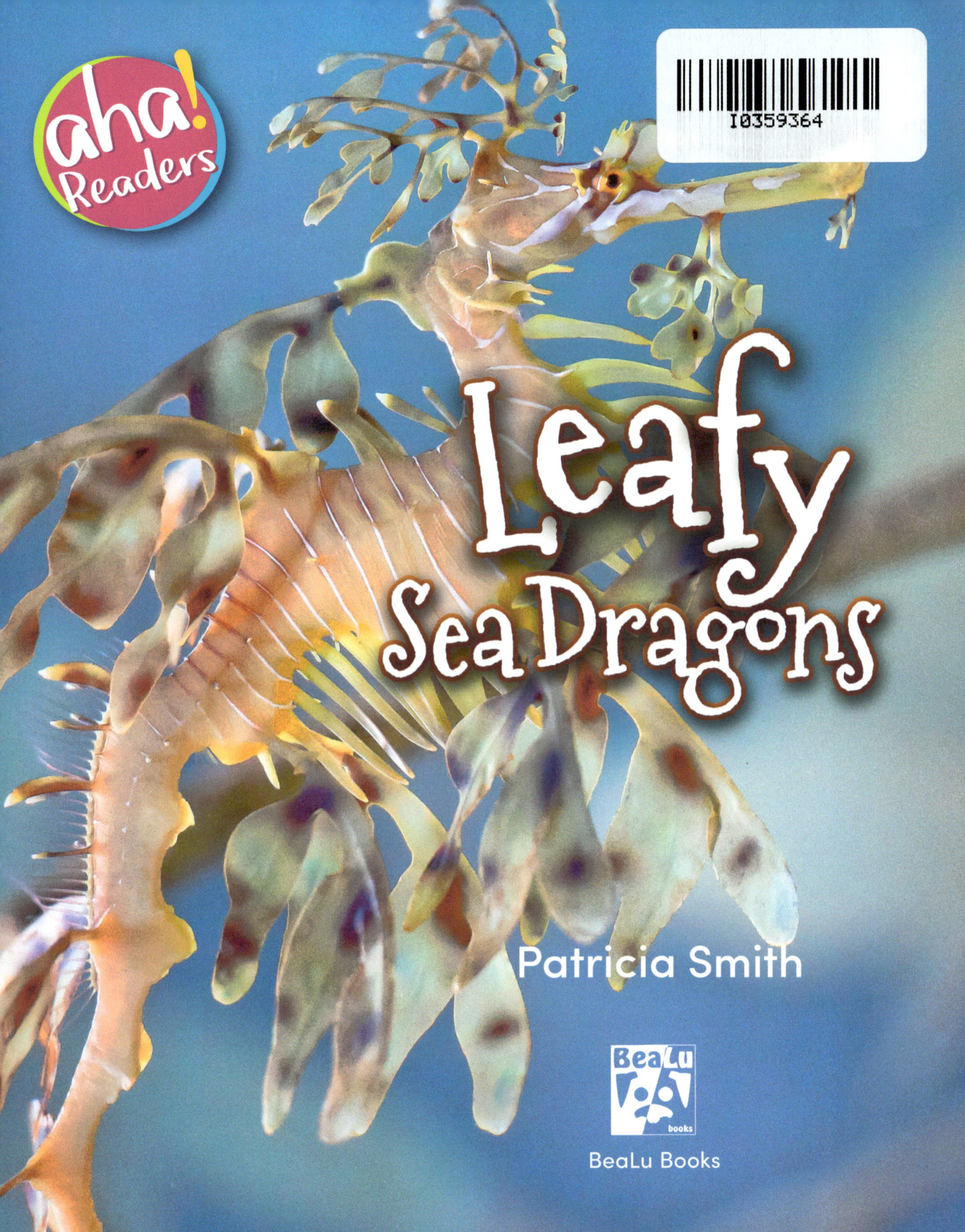

# Leafy Sea Dragons

Patricia Smith

BeaLu Books

## Dedication:

To my mom and dad, who gave me the world, and to Ryan who helped me to find my way back to it.

Copyright © 2020 by Patricia Smith

All rights reserved. No part of this publication may be reproduced in any form or by any electronic or mechanical means, including information storage and retrieval systems, without the express written permission from the publisher, except in the case of brief quotations embodied in critical articles or reviews. For information regarding permission, contact BeaLu Books.

ISBN Hardcover: 978-1-7341065-3-4
ISBN Paperback: 978-1-7333092-6-4

**Library of Congress Control Number:** 2019952459
Publisher's Cataloging-in-Publication Data is on file with the publisher.

Edited by: Luana K. Mitten
Book cover and interior design by Tara Raymo • creativelytara.com

Printed in the United States of America
October 2019

BeaLu Books
Tampa, Florida

www.BeaLuBooks.com

PHOTO CREDITS: Cover: © Ian D M Robertson; Page 1: © Kevin Ouellette; Page 3: © Jun Zhang; Pages 4-5: © AshtonEa; Page 7: © Zoe Esteban; Pages 8-9: © imageZebra; Page 10: © dikobraziy; Page 11: © goobafish; Page 13: © Kris Wiktor; Page 14-15: © Cyril Hou: Page 16: © ronnybas frimages; Page 17: © John Carnemolia, © superjoseph, © Aquarius Photography, © Tanya Puntti

Far beneath the surface of the ocean lives an amazing creature. It is called the leafy sea dragon, but it is not a sea horse or a dragon or a plant. A leafy sea dragon is a fish with bony plates and rings on its body. It is covered with growths that look like leaves.

The sea dragon can change colors and uses these leafy growths to blend in, or camouflage, with its environment. Sea dragons range in color from yellow to green to brown to deep red and may be hard to see when swimming among underwater plants and reefs. This helps the leafy sea dragon to protect itself from other fish.

Seaweed beds, reefs, and sea grass meadows off the coast of Australia are home to the leafy sea dragons. The creatures need warm salt water to live, and small ocean plants and animals to eat.

Sea dragons may travel many miles to find food in the winter, but scientists have discovered that they can find their way back to their original home. Since the sea dragon moves very slowly—only one mile in eight hours—getting home again may take a while.

Usually sea dragons live apart from other sea dragons, but in October to March of each year, they gather together. Then a male and female sea dragon will come together to mate. The female lays between 250 and 300 tiny eggs, which she places in the soft skin under the male's tail.

The male carries the eggs and holds them until they are ready to hatch. After about six weeks, the male releases the eggs a few at a time. Hours or days later, all of the baby sea dragons are swimming in the ocean by themselves.

A newly hatched sea dragon is very small. It is silver and black and has a very short snout. For a couple of days, it will stay near the area where it was hatched and will settle in shallow water. Its parents do not take care of it or of its brothers and sisters. Most of the newly hatched babies will be eaten by other fish and animals.

Those sea dragons that live to adulthood may grow to be up to fourteen inches long and can live from seven to ten years. In Australia, people believe that sea dragons bring good luck and many towns celebrate the amazing fish at festivals. The leafy sea dragon is the official aquatic emblem of South Australia.

Leafy sea dragons are protected because their survival is threatened. Today there are fewer sea dragons, and fewer kinds of sea dragons, than there once were. Pollution and loss of habitat are dangers to these fish.

Scientists are working to protect sea dragons by tracking their movements and recording the information into computer databases. The Australian government has passed laws to forbid divers from taking the fish. Many people want to help sea dragons to survive.

Sea dragons' habitat is in Australia, but they also live in aquariums around the world. People in many countries, including America, can visit leafy sea dragons at nearby aquariums and see these wonderful creatures up close.

# The Great Barrier Reef

*Australia's Great Barrier Reef, where leafy sea dragons live, is an amazing place. Stretching more than 1,800 miles (2,900 kilometers), it is made of living corals that grow in many shapes and colors on older, dead corals. The Great Barrier Reef is the world's largest reef system and is home to thousands of species of plants and animals, including sea stars, snakes, turtles, and many kinds of birds.*

*You can help to protect the Great Barrier Reef from the pollution and warmer ocean temperatures that threaten it. You can reuse things you buy and recycle materials like paper and plastic to help protect the environment. You can avoid buying jewelry or products made from ocean animals. Finally, you can work with organizations like the Coral Reef Alliance to protect the Great Barrier Reef and the plants and animals living there. Together with the scientists, leaders, and citizens of countries around the world, you can help to preserve a place unlike any other on Earth.*

17

## About the author:

Patricia Smith has family throughout Europe and spoke German as her first language. Patricia and her mother learned English together by watching Sesame Street, and by the time Patricia was three she could translate conversations for her German-speaking grandmother. Though she doesn't have children of her own, Patricia loves working with her students and taking care of five cats.

www.natureaustralia.org.au
(scroll to find the live Reef Cam)

www.coral.org

www.barrierreef.org

www.aims.gov.au

www.ingramcontent.com/pod-product-compliance
Lightning Source LLC
Chambersburg PA
CBHW041125070526
44584CB00003B/282